Probability

Outline and Activities

What's Included:

Outline of Probability Topics (3 pages)

Basic Questions by Topic (4 pages)

Basic Questions by Topic Answer Key (4 pages)

2 Probability Puzzles (Probability of Single & Compound Events) (4 pages)

6 Probability Worksheets each with 12 random questions (6 pages)

6 Probability Worksheets Answer Key (6 pages)

6 Short Answer Questions with Answer Key (4 pages)

Probability Outline

Probability - ***The Chance (or likelihood) of something (an event) happening***

1) Probability Basics

- Probability can be written as a ***fraction, decimal, or percentage***. (ex. ½, .50 or 50%)

- When writing as a fraction the top represents the **favorable outcomes** (what you want to occur) and the bottom represents all the **total possible outcomes**.

- For example, when flipping a coin and predicting ***"Heads."*** The Probability that heads will occur is ½. The "1" represents the favorable outcomes (***Heads***) and the "2" represents all the possible outcomes (***Heads or Tails***).

$$\frac{\textbf{Favorable Outcome}}{\textbf{Total Possible Outcomes}} \qquad \frac{\textbf{Getting Heads when flipping a coin}}{\textbf{Getting either Heads or Tails}}$$

- An **Event** is the action or the thing that is happening. For example, rolling a number cube, flipping a coin, and spinning a spinner are all examples of an event.

- ***P(Heads)*** is a way that a probability question could be asked. The capital ***"P"*** represents the word ***"Probability"*** and what ever is inside the parentheses ( ) represents the favorable outcome.

So, *P (Tails)* = **Probability of getting Tails** and *P (2)* = **Probability of getting a "2."**

2) Probability on a Number Line

- The Number line used in probability is very simple: ***0 to 1***
- **"0"** = Can't happen (It has a 0% chance of occurring)
- **"1"** = Guaranteed to happen (It has a 100% chance of occurring)

Examples:

- *Rolling a number cube (numbers 1 to 6) & getting a "7" is 0% chance = "0" on number line*
- *Rolling a number cube & getting a number less than 7 is 100%. chance = "1" on number line*

The closer an event is to "0" the less likely it is to happen.
The closer an event is to "1" the greater the chance it is to happen.

3) Theoretical Probability

- What we expect to happen based on the probability. For example, you have a ½ chance of flipping a coin and getting heads. So, if you flip a coin 50 times you would expect to get Heads 25 times (25 is half of 50).

- If you roll a Die 60 times the number "2" should occur 10 times (1/6 chance). Actually, each the numbers should occur 10 times.

) Theoretical v. Experimental

- ***Experimental Probability*** is the actual act of doing the experiment (example: flipping a coin 50 times or rolling a die 60 times) and recording the results.

- Many times the ***Theoretical*** and ***Experimental*** are different but *usually* similar.

Example: You flipped a coin 10 times. The Theoretical Probability says ***heads*** and ***tails*** should each occur 5 out of 10 times. When the experiment was conducted you had 4 heads and 6 tails.

How is the results of your experiment different from the theoretical probability? Heads occurred 1 time less than expected and Tails occurred 1 more than expected.

) Simple Event (single event) v. Compound Events

- ***Simple event*** is a single flip, roll, spin etc.
    - Example: P(rolling a number cube and getting a "2")
- ***Compound event*** is 2 or more flips, rolls, spins, etc.
    - Example: P(rolling a number cube and getting a "2" three straight times)

) Frequency Table (aka Sample Space), Organized List, & Tree Diagrams

- Ways to visualize the different combinations of events.

Below are 3 different ways to Model all the combinations of flipping two coins:

Frequency Table Example: Showing all the combinations of tossing two coins.

	HEADS *(H)*	TAILS *(T)*
HEADS (H)	H *H*	H *T*
TAILS (T)	T *H*	T *T*

Organized List Example: Showing all the combinations of tossing two coins.

Heads & Heads
Heads & Tails
Tails & Heads
Tails & Tails

Tree Diagram Example: Showing all the combinations of tossing two coins.

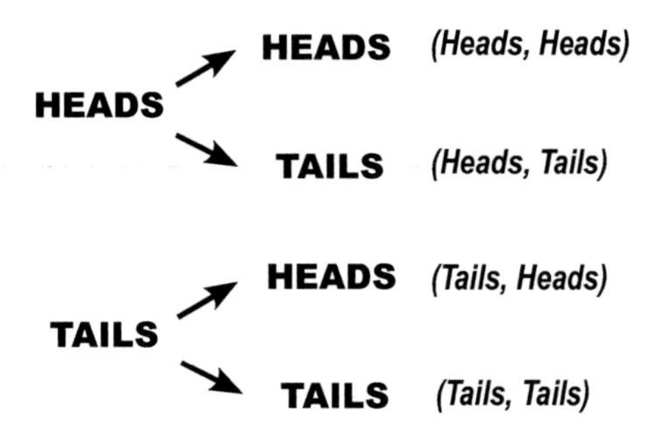

7) Math behind the number of combinations

- Each event has a fraction of a chance of happening. Multiply all the fractions representing each events chance of occurring. The answer is the probability of this combination of events occurring.

Examples:

Flipping a coin 2 times and getting heads both times: $\frac{1}{2} \times \frac{1}{2} = \frac{1}{4}$
Flipping a coin 3 times and getting tails all three times: $\frac{1}{2} \times \frac{1}{2} \times \frac{1}{2} = \frac{1}{8}$
Rollin a number cube (numbers 1-6) 2 times and getting a five both times: $\frac{1}{6} \times \frac{1}{6} = \frac{1}{36}$

8) Dependent v. Independent

Independent - When 2 or more events occur and they are not affected by the outcome of the other events

Example:
Flipping a coin is **Independent** - Each time you flip a coin you have a ½ chance of getting tails. Regardless of how many times you flip a coin the probability of getting tails is always ½.

Dependent- The probability of an event is changed by the results of previous event(s).

Example:
A jar is filled with a total of ten marbles. 3 marbles are red and 7 are blue. If a person randomly selects a marble they have a $\frac{3}{10}$ chance of getting a red marble. If that person keeps the red marble the odds for the next person getting a red marble is now $\frac{2}{9}$.

The second person's probability of getting a red marble is dependent on what happens with the first selection. If the first person had selected a blue marble the second person would actually have a better probability of getting a red ($\frac{3}{9}$ or $\frac{1}{3}$). It's possible that the 4th person picking a marble might have a 0% chance of getting a red marble if the first 3 people each select a red marble (Assuming that each person is not returning the marbles selected).

9) Combinations (similar to Dependent Probability)

- This is the number of different things (arrangements) that can occur. The number of options changes each time a new event is added.

Example: You have 3 baseball hats (LA, NY, & SF). When you go to select the first hat you have three options, the second hat you will only have 2 options and the last hat will be the only hat left. So...

3 x 2 x 1 = 6 (the total of different ways to arrange the three hats)
The 6 combos: LNS, LSN, NLS, NSL, SLN, SNL

The probability that someone could predict the arrangement is ***1 out of 6 or ⅙***

10) Uniformed Probability

- When each event has the same probability of occurring.

2 Examples of Uniformed Probability: **2 Examples of *NOT* having Uniformed Probability**

#1 Basic Probability

- What is the probability of guessing the correct answer on a True and False Question. __________
- What is the probability of rolling a #3 on a number cube (numbered 1 to 6)? __________
- What is the probability of rolling an even number on a number cube? __________
- What is the probability of selecting a "Diamond" from a deck of cards? __________
- You have 7 red marbles, 3 green marbles, and 5 yellow marbles.

  a) P(picking red) = ______ b) P(picking green) =______ c) P(picking yellow)= ______

  d) P(not picking red)= _____ e) P(Blue) _____

- What is the probability that a random day during the week ends in the letter "y"? __________

#2 Number Line

'rite the number of the problem on the number line below that best represents the probability of that question.

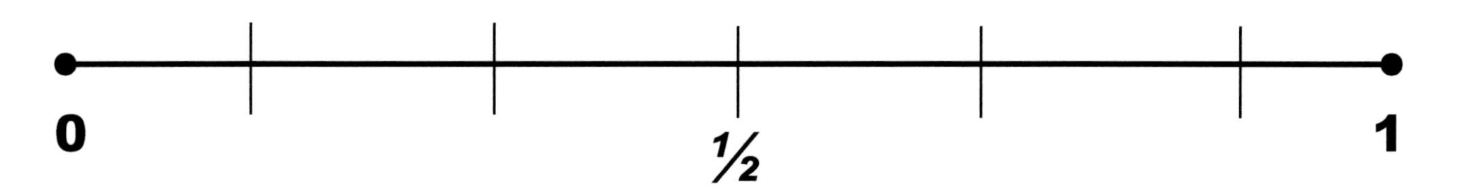

- The probability of guessing the correct answer on a coin flip.
- The probability of rolling a #5 on a number cube (numbered 1 to 6)?
- The probability of rolling a number greater than "2" on a number cube?
- The probability of winning the Mega Millions lottery?
- You have 7 red marbles, 1 green marbles, and 12 yellow marbles. P(of not getting a green marble)
- The probability that you will roll a number greater than 6 when rolling a number cube (numbered 1 to 6)?

#3 Theoretical Probability

1) If you flip a coin 200 times, how many times would you expect "heads" to occur? ____________________

2) If you rolled a number cube (numbered 1 to 6) 120 times, how many times should the number "4" occur?

3) You have a spinner divided into 4 equal sections (A,B,C,D). If you spin the spinner 44 times how many times should the letter "A" occur?

4) You have a spinner divided into 3 equal sections (A,B,C). If you spin the spinner 60 times, how many times should the letter "B" occur?

5) You have a jar with 2 green, 7 red and 11 blue marbles. If 200 kids reached in a selected a marble (and returned how many kids should have selected Red? ________ Green? _________ Blue? _________ Purple? _________

#4 Theoretical vs. Experimental Probability

1) You flipped a coin 140 times. Heads occurred 80 times. How does this compare to the Theoretical Probability?

2) If you rolled a number cube (numbered 1 to 6) 66 times. The number "4" occurred 11 times? How does this compare to the Theoretical Probability?

3) You have a spinner divided into 4 equal sections (A,B,C,D). After spinning the spinner 84 times the letter "B" occurred 25 times? How does this compare to the Theoretical Probability?

4) You have a spinner divided into 3 equal sections (A,B,C). After spinning the spinner 240 times, the letter "C" occurred 60 times? How does this compare to the Theoretical Probability?

5) You have a jar with 2 green, 7 red and 11 blue marbles. 500 kids selected a marble (then returned to the jar). The Green marble was selected 100 times. How does this compare to the Theoretical Probability?

#5 Compound Events (independent)

) You flipped a coin 2 times. What is the probability of getting tails both times?

) You are taking a True and false quiz. There are 5 questions. What is the probability of correctly guessing all five uestions?

) What is the probability of rolling a number cube twice and getting a "5" on both rolls?

) You flipped a coin 4 times. What is the probability of getting heads all 4 times?

) You have a spinner with 4 equal sections (A,B,C,D) and a spinner with 3 equal sections (1,2,3). What is the probaility of spinning the two spinners and getting the combination of "A2"?

) You have a jar with 9 marbles (3 red, 5 orange, and 1 blue). What is the probability of selecting a blue marble 2 traight times (returning the marble after the first selection) in a row?

#6 Modeling possible outcomes of Compound Events

) You flipped a coin 2 times. Model with a Frequency Table (sample space), Organized List and Tree Diagram

) You are taking a True and False quiz. There are 3 questions. Use a Tree Diagram to show all the combinations.

) You have a spinner with 4 equal sections (A,B,C,D) and a spinner with 3 equal sections (1,2,3). Write an orgaized list of all the different combinations.

) A menu has three different drinks (coke, pepsi, and milk) and three different sandwiches (fish, ham, and sloppy oe). Create a Tree Diagram and a Frequency Table (sample space) of all the different combinations.

#7 Compound Events (Independent & Dependent)

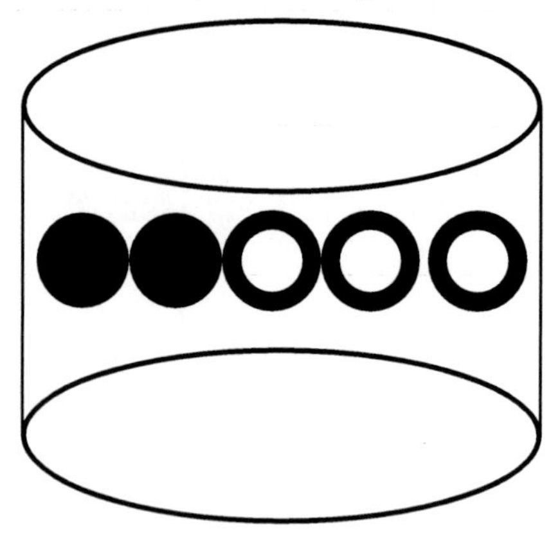

1) P(picking 2 straight black marbles *replacing* after selecting)

2) P(picking 2 straight black marbles *without* replacing)

3) P(picking 2 straight open circle marbles *replacing* after selecting)

4) P(picking 2 straight open circle marbles *without* replacing)

5) P(picking 3 straight open circle marbles *without* replacing)

6) P(picking 3 straight black marbles *without* replacing)

#8 Amount of Combinations

1) You have 5 different baseball hats (Ex. LA, NY, SD, TX, & SF) . How many different ways can they be arranged? (Ex. NY, LA, SD, TX, & SF)

2) You have 4 different candy bars (Hershey, Snickers, Pay Day and Reese's). How many different ways could you eat all candy bars. For example, eating Pay Day first followed by Hershey, Snickers and Reese's would be one w

3) You go to a local diner. The dinner special is: One Entree, One side and a drink. This restaurant has 4 Entrees, sides and 5 drinks to select from. How many different combinations are possible?

4) The University of Oregon is know for their unique football uniforms. Oregon has 4 different helmets, 6 differen jersey tops and 5 different bottoms. How many combinations are possible?

Answer Key

#1 Basic Probability

1) What is the probability of guessing the correct answer on a True and False Question. $1/2$

2) What is the probability of rolling a #3 on a number cube (numbered 1 to 6)? $1/6$

3) What is the probability of rolling an even number on a number cube? $1/2$

4) What is the probability of selecting a "Diamond" from a deck of cards? $1/4$

5) You have 7 red marbles, 3 green marbles, and 5 yellow marbles.

a) P(red) = $7/15$ b) P(green) = $3/15 = 1/5$ c) P(yellow)= $5/15 = 1/3$ d) P(not red)= $8/15$ e) P(Blue) $0/15 = 0$

6) What is the probability that a random day during the week ends in the letter "y"? $7/7 = 1$

#2 Number Line

Write the number of the problem on the number line below that best represents the probability of that question.

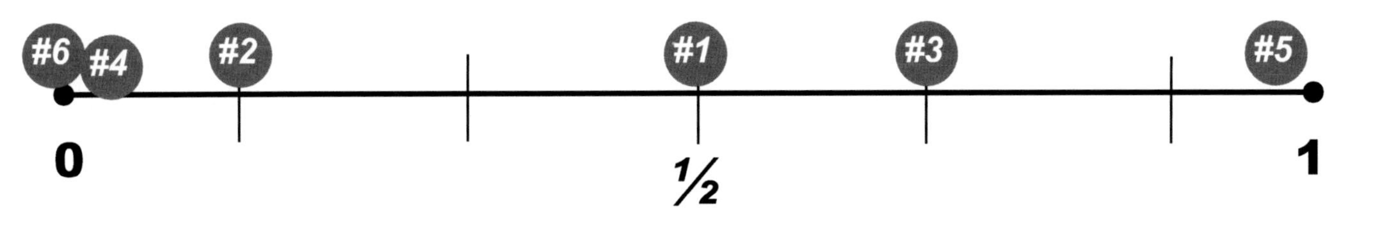

1) The probability of guessing the correct answer on a coin flip. $1/2$

2) The probability of rolling a #5 on a number cube (numbered 1 to 6)? $1/6$

3) The probability of rolling a number greater than "2" on a number cube? $4/6 = 2/3$

4) The probability of winning the Mega Millions lottery? **Near Zero**

5) You have 7 red marbles, 1 green marbles, and 12 yellow marbles. P(of not getting a green marble) $19/20$
- Near One

6) The probability that you will roll a number greater than 6 when rolling a number cube (numbered 1 to 6)? $0/6$

Answer Key

#3 Theoretical Probability

1) If you flip a coin 200 times, how many times would you expect "heads" to occur? *100*

2) If you rolled a number cube (numbered 1 to 6) 120 times, how many times should the number "4" occur?
20

3) You have a spinner divided into 4 equal sections (A,B,C,D). If you spin the spinner 44 times how many times should the letter "A" occur?
11

4) You have a spinner divided into 3 equal sections (A,B,C). If you spin the spinner 60 times, how many times should the letter "B" occur?
20

5) You have a jar with 2 green, 7 red and 11 blue marbles. If 200 kids reached in a selected a marble (and returnec how many kids should have selected Red? *70* Green? *20* Blue? *110* Purple? *0*

#4 Theoretical vs. Experimental Probability

1) You flipped a coin 140 times. Heads occurred 80 times. How does this compare to the Theoretical Probability?

Theoretical Probability = 70 out of 140. Experimental Probability = 70 out of 140. Heads occurred 10 more than expected

2) If you rolled a number cube (numbered 1 to 6) 66 times. The number "4" occurred 11 times? How does this com pare to the Theoretical Probability?

Theoretical Probability = 11 out of 66. Experimental Probability = 11 out of 66. Same.

3) You have a spinner divided into 4 equal sections (A,B,C,D). After spinning the spinner 84 times the letter "B" occurred 25 times? How does this compare to the Theoretical Probability?

Theoretical Probability = 21 out of 84. Experimental Probability = 25 out of 84. "B" occurred 4 more than expected

4) You have a spinner divided into 3 equal sections (A,B,C). After spinning the spinner 240 times, the letter "C" occurred 60 times? How does this compare to the Theoretical Probability?

Theoretical Probability = 80 out of 240. Experimental Probability = 60 out of 240. "B" occurred 20 less than expected

5) You have a jar with 2 green, 7 red and 11 blue marbles. 500 kids selected a marble (then returned to the jar). Th Green marble was selected 100 times. How does this compare to the Theoretical Probability?

Theoretical Probability = 50 out of 500. Experimental Probability = 100 out of 500. Green occurred 50 more than expectec

#5 Compound Events (independent)

You flipped a coin 2 times. What is the probability of getting tails both times?

$$\frac{1}{2} \times \frac{1}{2} = \frac{1}{4}$$

You are taking a True and false quiz. There are 5 questions. What is the probability of correctly guessing all five uestions ?

$$\frac{1}{2} \times \frac{1}{2} \times \frac{1}{2} \times \frac{1}{2} \times \frac{1}{2} = \frac{1}{32}$$

What is the probability of rolling a number cube twice and getting a "5" on both rolls?

$$\frac{1}{6} \times \frac{1}{6} = \frac{1}{36}$$

You flipped a coin 4 times. What is the probability of getting heads all 4 times?

$$\frac{1}{2} \times \frac{1}{2} \times \frac{1}{2} \times \frac{1}{2} = \frac{1}{16}$$

You have a spinner with 4 equal sections (A,B,C,D) and a spinner with 3 equal sections (1,2,3). What is the proba-lity of spinning the two spinners and getting the combination of "A2"?

$$\frac{1}{4} \times \frac{1}{3} = \frac{1}{12}$$

You have a jar with 9 marbles (3 red, 5 orange, and 1 blue). What is the probability of selecting a blue marble 2 raight times (returning the marble after the first selection) in a row?

$$\frac{1}{9} \times \frac{1}{9} = \frac{1}{81}$$

#6 Modeling possible outcomes of Compound Events

You flipped a coin 2 times. Model with a sample space (box), Organized List and Tree Diagram

HEADS, TAILS
HEADS, HEADS
TAILS, HEADS
TAILS, TAILS

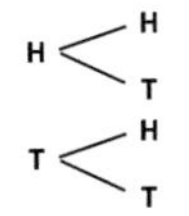

	H	T
H	HH	HT
T	TH	TT

You are taking a True and False quiz. There are 3 questions. Use a Tree Diagram to show all the combinations.

T — T — T, F
T — F — T, F
F — T — T, F
F — F — T, F

You have a spinner with 4 equal sections (A,B,C,D) and a spinner with 3 equal sections (1,2,3). Write an orga-ized list of all the different combinations.

A1, A2, A3
B1, B2, B3
C1, C2, C3
D1, D2, D3

A menu has three different drinks (coke, pepsi, and milk) and three different sandwiches (fish, ham, and sloppy e). Create a tree diagram and a sample space.

	C	P	M
F	FC	FP	FM
H	HC	HP	HM
S	SC	SP	SM

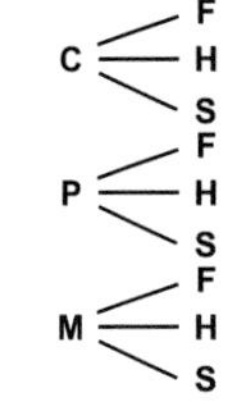

Answer Key

#7 Compound Events (Independent & Dependent)

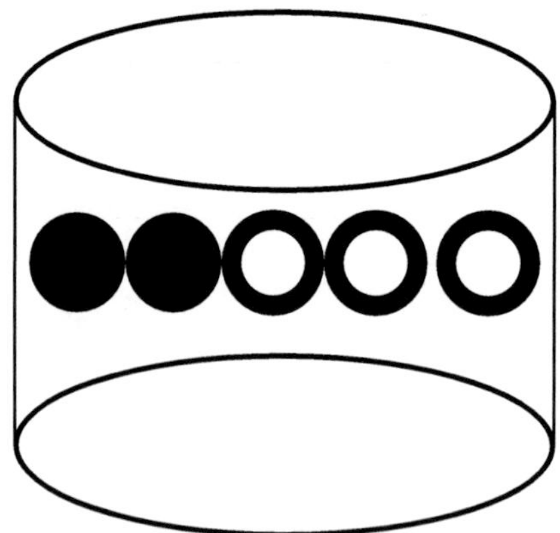

1) P(picking 2 straight black marbles *replacing* after selecting) $\frac{2}{5} \times \frac{2}{5} = \frac{4}{25}$

2) P(picking 2 straight black marbles *without* replacing) $\frac{2}{5} \times \frac{1}{4} = \frac{2}{20} = \frac{1}{10}$

3) P(picking 2 straight open circle marbles *replacing* after selecting) $\frac{3}{5} \times \frac{3}{5} = \frac{9}{25}$

4) P(picking 2 straight open circle marbles *without* replacing) $\frac{3}{5} \times \frac{2}{4} = \frac{6}{20} = \frac{3}{10}$

5) P(3 straight open circle marbles *without* replacing) $\frac{3}{5} \times \frac{2}{4} \times \frac{1}{3} = \frac{6}{60} = \frac{1}{10}$

6) P(3 straight black marbles *without* replacing) $\frac{2}{5} \times \frac{1}{4} \times \frac{0}{3} = \frac{0}{60} = 0$

#8 Amount of Combinations

1) You have 5 different baseball hats (Ex. LA, NY, SD, TX, & SF) . How many different ways can they be arranged? (Ex. NY, LA, SD, TX, & SF)

$5 \times 4 \times 3 \times 2 \times 1$ = 120 ways

2) You have 4 different candy bars (Hershey, Snickers, Pay Day and Reese's). How many different ways could you eat all candy bars. For example, eating Pay Day first followed by Hershey, Snickers and Reese's would be one wa

$4 \times 3 \times 2 \times 1$ = 24 ways

3) You go to a local diner. The dinner special is: One Entree, One side and a drink. This restaurant has 4 Entrees, sides and 5 drinks to select from. How many different combinations are possible?

$4 \times 3 \times 5$ = 60 combinations

4) The University of Oregon is know for their unique football uniforms. Oregon has 4 different helmets, 6 different jersey tops and 5 different bottoms. How many combinations are possible?

$4 \times 6 \times 5$ = 120 combinations

2 Probability Puzzles

20 puzzle pieces each

Over 20 questions for each puzzle

Puzzle #1: Simple Events
Puzzle #2: Compound Events

Solution: ***"MY SALTED CARAMEL MOCHA"*** **Topic:**

Read from the bottom up.

Right to left, then left to right etc. (follow the arrows)

Probability of a Simple (single) Event (Some simplification needed)

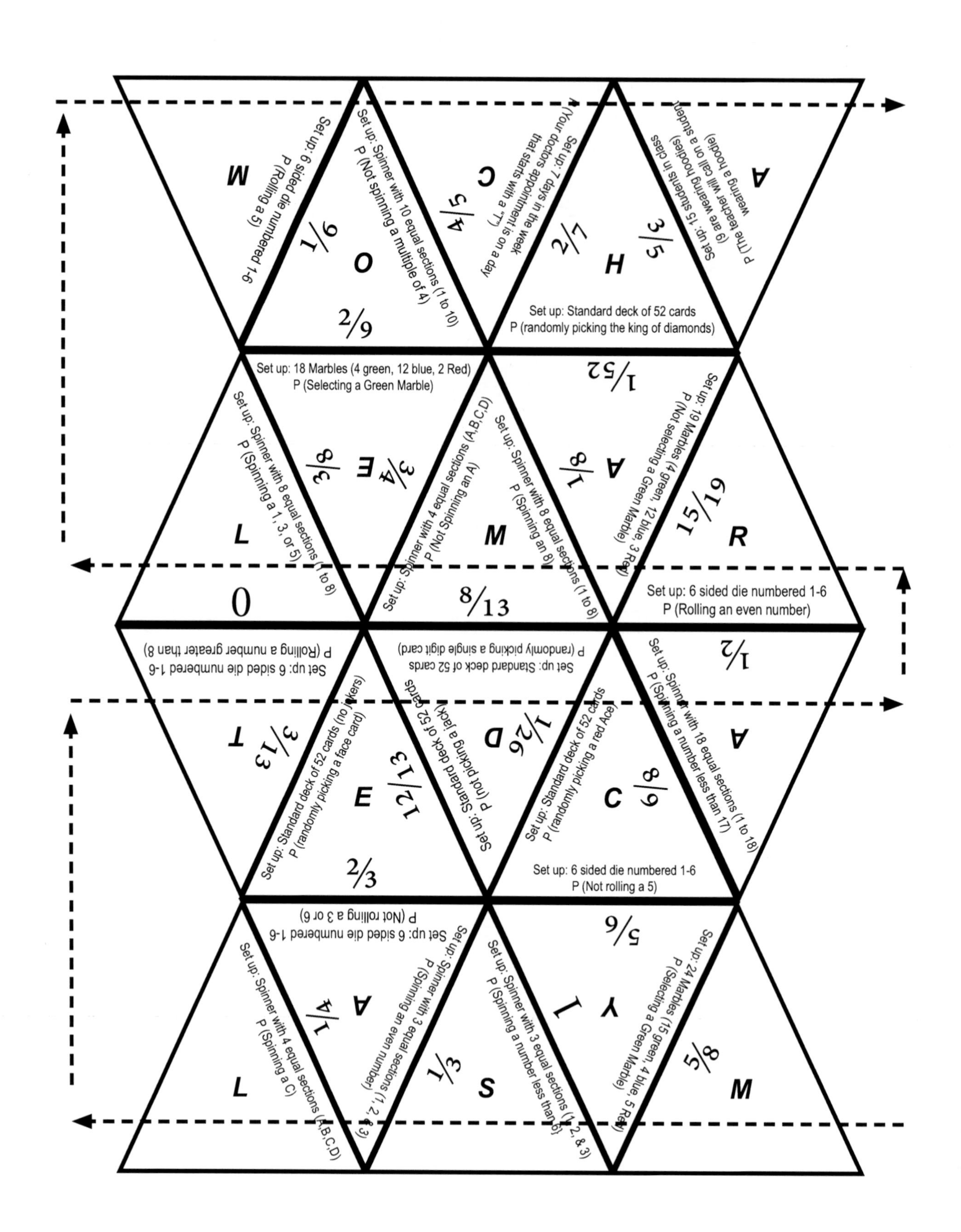

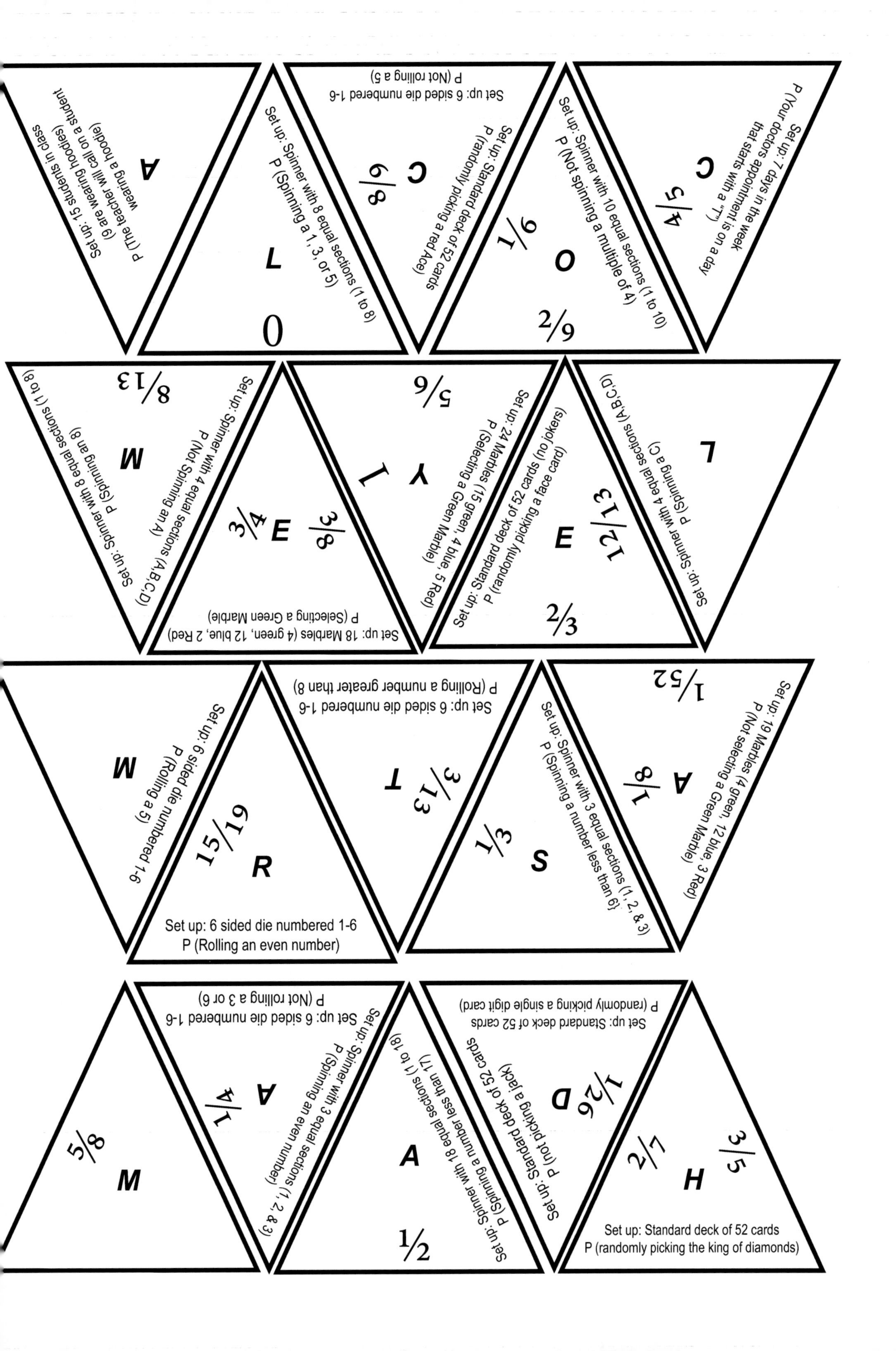
A
Set up: 15 students in class
(9 are wearing hoodies)
P (The teacher will call on a student wearing a hoodie)
L
Set up: Spinner with 8 equal sections (1 to 8)
P (Spinning a 1, 3, or 5)
0
Set up: 6 sided die numbered 1-6
P (Not rolling a 5)
C
6/8
Set up: Standard deck of 52 cards
P (randomly picking a red Ace)
9/1
O
2/9
Set up: Spinner with 10 equal sections (1 to 10)
P (Not spinning a multiple of 4)
C
4/5
Set up: 7 days in the week
P (Your doctors appointment is on a day that starts with a "T")
M
8/13
Set up: Spinner with 8 equal sections (1 to 8)
P (Spinning an 8)
Set up: Spinner with 4 equal sections (A,B,C,D)
P (Not Spinning an A)
3/4
E
3/8
Set up: 18 Marbles (4 green, 12 blue, 2 Red)
P (Selecting a Green Marble)
5/6
1
Y
Set up: 24 Marbles (15 green, 4 blue, 5 Red)
P (Selecting a Green Marble)
Set up: Standard deck of 52 cards (no jokers)
P (randomly picking a face card)
E
12/13
2/3
L
Set up: Spinner with 4 equal sections (A,B,C,D)
P (Spinning a C)
M
Set up: 6 sided die numbered 1-6
P (Rolling a 5)
15/19
R
Set up: 6 sided die numbered 1-6
P (Rolling an even number)
Set up: 6 sided die numbered 1-6
P (Rolling a number greater than 8)
3/13
T
1/3
S
Set up: Spinner with 3 equal sections (1, 2, & 3)
P (Spinning a number less than 6)
1/52
A
1/8
Set up: 19 Marbles (4 green, 12 blue, 3 Red)
P (Not selecting a Green Marble)
5/8
M
1/4
A
Set up: 6 sided die numbered 1-6
P (Not rolling a 3 or 6)
Set up: Spinner with 3 equal sections (1, 2, & 3)
P (Spinning an even number)
A
1/2
Set up: Spinner with 18 equal sections (1 to 18)
P (Spinning a number less than 17)
D
1/26
Set up: Standard deck of 52 cards
P (not picking a jack)
Set up: Standard deck of 52 cards
P (randomly picking a single digit card)
2/7
3/5
H
Set up: Standard deck of 52 cards
P (randomly picking the king of diamonds)

Solution: Top: *"Buccaneers"* Bottom: *"Catamounts"*

Topic:

Probability of a Compound (2+) Event (Some simplification needed)

Start in the upper right and read counter clockwise. The top 10 triangles spell out "Buccaneers" and the bottom 10 triangles spell out "Catamounts."

Some of the repeat letters are in different fonts to help check answers quicker.

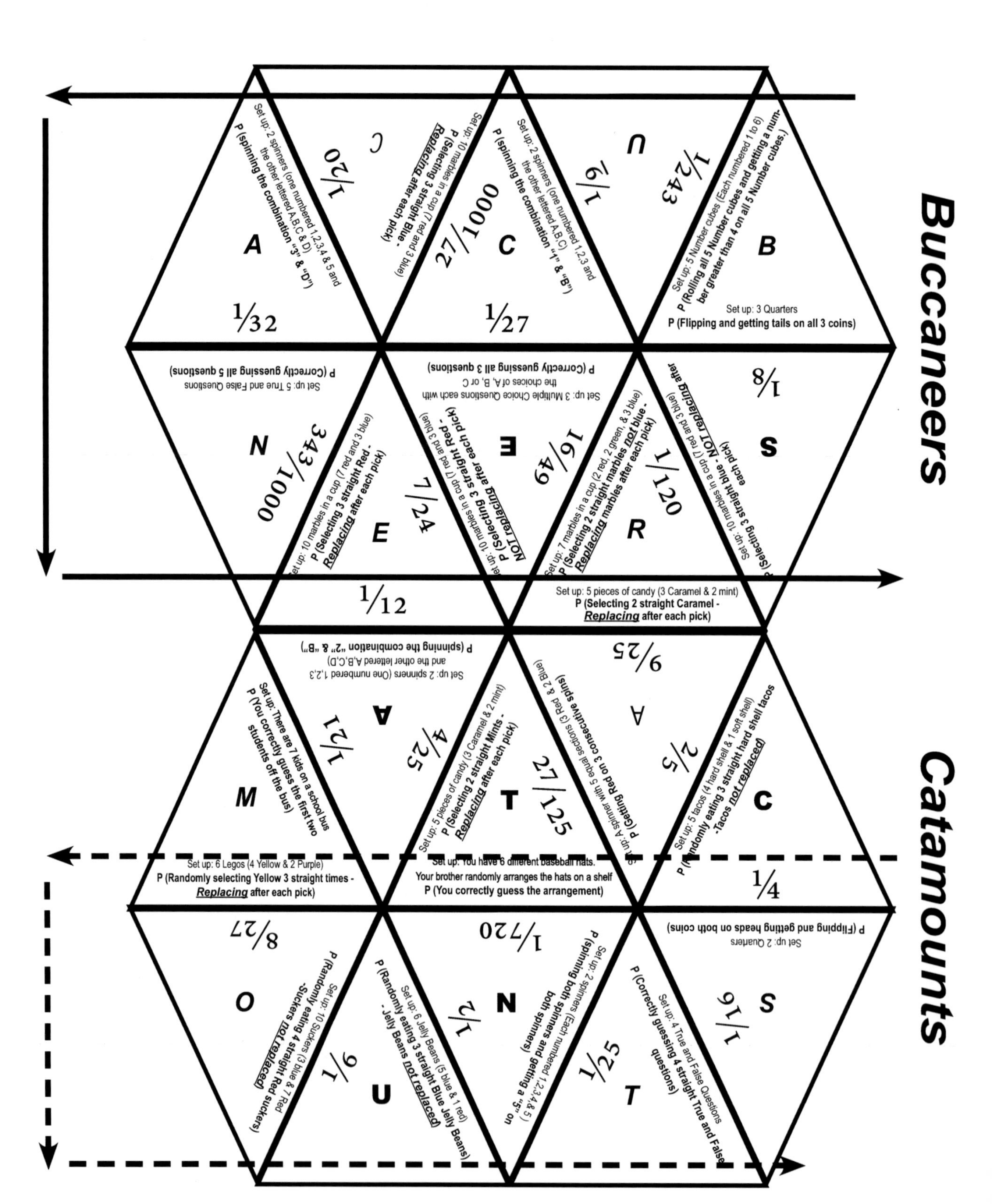

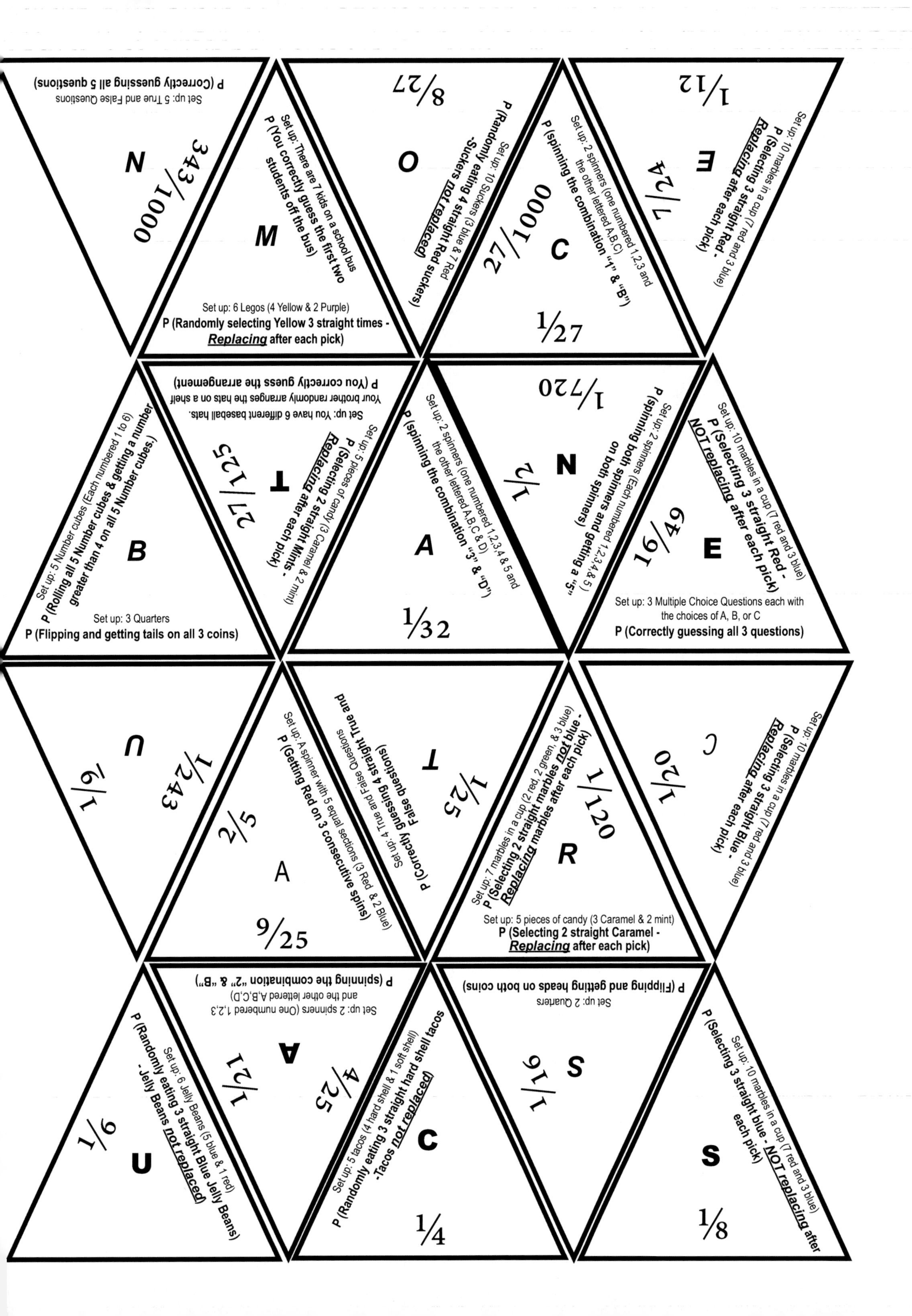
Set up: 5 True and False Questions
P (Correctly guessing all 5 questions)
343/1000
N
Set up: There are 7 kids on a school bus
P (You correctly guess the first two students off the bus)
M
Set up: 6 Legos (4 Yellow & 2 Purple)
P (Randomly selecting Yellow 3 straight times - Replacing after each pick)
8/27
Set up: 10 Suckers (3 blue & 7 Red
P (Randomly eating 4 straight Red suckers) -Suckers not replaced)
O
27/1000
Set up: 2 spinners (one numbered 1,2,3 and the other lettered A,B,C)
P (spinning the combination "1" & "B")
C
1/27
1/12
Set up: 10 marbles in a cup (7 red and 3 blue)
P (Selecting 3 straight Red - Replacing after each pick)
E
7/24
Set up: 5 Number cubes (Each numbered 1 to 6)
P (Rolling all 5 Number cubes & getting a number greater than 4 on all 5 Number cubes.)
B
Set up: 3 Quarters
P (Flipping and getting tails on all 3 coins)
Set up: You have 6 different baseball hats. Your brother randomly arranges the hats on a shelf
P (You correctly guess the arrangement)
27/125
T
Set up: 5 pieces of candy (3 Caramel & 2 mint)
P (Selecting 2 straight Mints - Replacing after each pick)
Set up: 2 spinners (one numbered 1,2,3,4 & 5 and the other lettered A,B,C & D)
P (spinning the combination "3" & "D")
A
1/32
1/2
1/720
Set up: 2 spinners (Each numbered 1,2,3,4 & 5)
P (spinning both spinners and getting a "5" on both spinners)
N
Set up: 10 marbles in a cup (7 red and 3 blue)
P (Selecting 3 straight Red - NOT replacing after each pick)
16/49
E
Set up: 3 Multiple Choice Questions each with the choices of A, B, or C
P (Correctly guessing all 3 questions)
U
1/6
1/243
2/5
A
Set up: A spinner with 5 equal sections (3 Red & 2 Blue)
P (Getting Red on 3 consecutive spins)
9/25
Set up: 4 True and False Questions
P (Correctly guessing 4 straight True and False questions)
T
1/25
1/120
Set up: 7 marbles in a cup (2 red, 2 green, & 3 blue)
P (Selecting 2 straight marbles not blue - Replacing marbles after each pick)
R
Set up: 5 pieces of candy (3 Caramel & 2 mint)
P (Selecting 2 straight Caramel - Replacing after each pick)
C
1/20
Set up: 10 marbles in a cup (7 red and 3 blue)
P (Selecting 3 straight Blue - Replacing after each pick)
Set up: 6 Jelly Beans (5 blue & 1 red)
P (Randomly eating 3 straight Blue Jelly Beans - Jelly Beans not replaced)
1/6
U
Set up: 2 spinners (One numbered 1,2,3 and the other lettered A,B,C,D)
P (spinning the combination "2" & "B")
A
1/12
4/25
Set up: 5 tacos (4 hard shell & 1 soft shell)
P (Randomly eating 3 straight hard shell tacos -Tacos not replaced)
C
1/4
Set up: 2 Quarters
P (Flipping and getting heads on both coins)
S
1/16
Set up: 10 marbles in a cup (7 red and 3 blue)
P (Selecting 3 straight blue - NOT replacing after each pick)
S
1/8

6 Probability Worksheets

12 questions each

A random collection of different types of probability questions

ame: ________________________________ Period: ____ ***Worksheet: #1***

1)

1	3	2
2	1	3
2	3	1

Archie can select any spot on the board. What is the probability he will select a 3? Is this an example of uniform probability?

2) Ms. Fogler surveyed 30 7th grade students about their favorite colors. The following are the results: Blue (8), Red (7), Purple (10), and Yellow (5). If there are 180 students in the 7th grade, make a prediction (based on survey) how many would select each color.

Blue: _____

Red: ______

Purple: ______

Yellow: _______

3) A spinner is divided into 5 equal sections. Two sections are blue, one green, one yellow and one red.

What is the probability of ***NOT*** landing on red when spinning?

4) What is the probability that you could correctly guess the first four questions correct of a true & false quiz?

5) There are 30 students in your 5th period science class. Sixteen are girls. During the class, the office called for a student. The probability that the student was a girl would fall where on a number line?

a) zero

b) near zero

c) near middle

d) near one

e) one

6) A box contains 4 jelly filled and 6 cake doughnuts. What is the probability that the first 3 doughnuts eaten will be jelly filled?

7)

Tim spins the spinner above 20 times. These are his results:

A = 7	B = 5
C = 4	D = 4

How are these results different than the theoretical probability?

8) What is the probability of rolling a "1" three straight times with a six sided number cube?

') You are playing a game with the two spinners below. 'o win you must correctly select the letter and number.

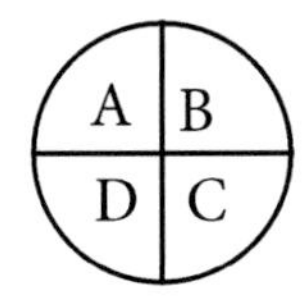

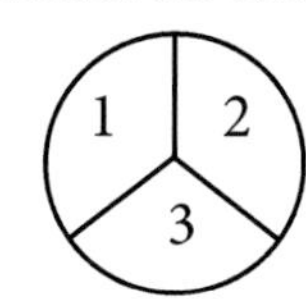

Vhat is the probability of spinning the combination of "A" nd "2"?

10) A restaurant offers 4 types of sandwiches (fish, hot dog, hamburger, grilled cheese) and three side dishes (fries, cole slaw, and onion rings). Create an organized list of all the possible outcomes if you order one sandwich and one side dish.

1) Is this an example of a uniformed probability model? xplain.

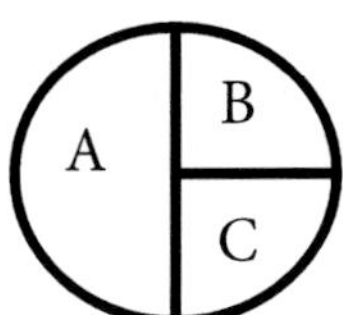

12) You flip a coin and roll a number cube (1 to 6). What is the probability that you will get the combination of tails and an even number?

Name: ______________________________ **Period:** ____ ***Worksheet:***

1) Mrs. Johnson's candy jar has 15 pieces of candy (7 orange, 5 grape, and 3 strawberry). Mrs. Johnson is letting four students select a piece of candy which they get to keep. What is the probability that each of the four pieces of candy will be strawberry?

2) Below is a different type of dart board for a new game. Is this an example of uniform probability? Explain.

B / C	A
A	D

3) Tim is going to put the following 6 marbles into a ja
(a) What is the probability of drawing a white marble?
(b) Is this an example of uniformed probability?

4) A jar contains 7 blue, 5 yellow, 7 white and 15 red marbles. Are you more likely to pull out a blue or a yellow marble? Explain

5) There are 20 jelly beans in a bag (5 yellow, 4 brown, 3 orange & 8 blue). What is the probability that the first three jelly beans you randomly eat will all be orange?

6) Edward has two hats (visor & baseball) and three t-shirts (green, red, & yellow). Create a Frequency table showing all the possible combinations.

7) You are rolling an 8 sided die (numbers 1 to 8). What is the probability of rolling a number 7 or larger?

8) You are going to do two separate events. The first event is flipping a coin. The second event is spinning a spinner with three equal sections (Red, Blue, & Green). Put the probability of each event occurring in order from least likely to most likely. **a) Getting heads, b) Spinning a green, c) Spinning an orange, d) getting either heads or tails.**

9) There are 4 multiple choice questions (each with 4 choices - only one correct). You had to guess on all four of these questions. What is the probability that you were able to guess all four questions correctly?

10) A jar contains 7 blue, 5 purple, 3 white and 15 red marbles. What is the probably of selecting a blue marble?

11) You are going to do two separate events. First, you are going to roll a number cube with sides numbered 1 to 6. Second, you are going to flip a coin. Put these four occurrences in order from least likely to most likely to happen on the number line. (use the letters A - D)

A) Getting Heads B) Getting Heads or Tails
C) Getting a "1" D) Not Getting a "1"

0 ½ 1

12) You have a six sided number cube (numbered 1 to 6). What is the probability of rolling an even number?

ame: ________________________________ **Period:** ____ ***Worksheet: #3***

1) You flip a coin and spin a spinner with four equal sections (sides A, B, C, D). What is probability of flipping heads and spinning "B" ? You can represent the probability as either a fraction, decimal, or percent.	2) Of the 20 students in Mr. Karp's science class, four are boys. If the office calls for a student, what are the chances it will be a girl? A) Impossible B) Unlikely C) Likely D) Certain	3) You have a spinner with three equal sections (A, B, C). Predict the number of times the spinner will land on "A" if you spin 100 times.	4) There are 7 boys in a class of 15. What is the probability of the teacher selecting a girl to answer the first question?

5) You are rolling a number cube (numbers 1 to 6) fifty times. Predict how many times a number less than four will occur. Justify your answer.	6) You have a spinner with four equal sections numbered 1 to 4. What is the probability of spinning a number greater than 3?	7) Jenny has 3 shirts (green, purple, red) and 3 shorts (tan, black, khaki) to pick out an outfit for school. Create a tree diagram showing all possible combinations.	8) A teacher needs to select a boy & girl to go to the office. There are three boys (Tim, Sam, Jack) and two girls (Mary, Valerie). Create an Organized List of all the outcomes.

9) Represent the probability of getting heads, when flipping a coin two times in a row as a decimal, fraction, and percent.	10) You are rolling a number cube (with the numbers: 1, 2, 3, 4, 5, 6) 120 times. Predict how many times the number 4 will occur. (theoretical probability).	11) You are flipping a coin 44 times. Based on the Theoretical Probability, predict the number of times tails will occur.	12) There are 13 red marbles, 5 green marbles and 2 blue marbles in a jar. Represent the probability of selecting a red marble as a fraction, decimal, and percent.

Name: ______________________________ **Period:** ____ ***Worksheet:***

1) If you flip a coin three times what is the probability of getting heads all three times? (write as a fraction)

2) You have a spinner with four equal sections (A, B, C, D). If you spin the spinner 80 times, how many times would you expect the "B" to occur?

3) There are 30 major league baseball teams. One is from Canada (Toronto Blue Jays). If each team had equal talent, how would you describe Toronto's chance of winning the World Series next year? (circle)

Impossible

Unlikely

Likely

Certain

4) If you flip a coin and roll a number cube (1 to 6) what is the probability you will get the combination of "Heads" and the number "5"?

5) Is this an example of uniform probability model?

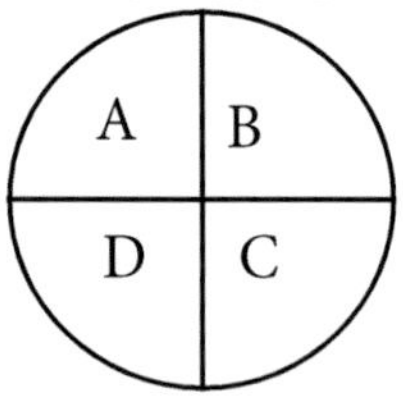

6) There are 15 blue cars and 5 red cars in the parking lot. Based on this sample, predict how many of the next 8 cars will be blue and red.

7) At the local coffee shop there are 3 sizes (S, M, L) and 4 drinks (Espresso, Frappuccino, Coffee, & Tea). Make a tree diagram of the different possible outcomes.

8) The following are the results of flipping a coin 20 times.

T H H H H T H H T T
T H H H T H H H T H

What is the experimental probability of tails?

9) You are one of the 200,000,000 people playing the lottery Friday. Your probability of winning would be located where on a number line?

a) Near zero

b) Near the middle

c) Near 1

10) There are 3 boys & 2 girls in Mr. Smith's class. There is 1 boy & 4 girls in Mrs. Greene's class. What is the probability a boy from Mr. Smith's class and Mrs. Greene's class will be selected as class representatives? Write as a fraction or percent.

11) There are 11 players on a football team. Five of the players are offensive linemen. If there is a penalty on the offense what is the probability it is on one of the offensive linemen?

12) You are playing cards with a standard deck of cards (52). In the first 40 games, how many times would you predict a face card (Jack, Queen, or King) would be the first card delt?

ame: ________________________________ **Period:** ____ ***Worksheet: #5***

1) A local radio station is having a contest to win a car. To win, a person must spin a spinner and land on a ***BLUE*** space. Each contestant has the choice of two spinners (below)to select from. Which spinner gives the contestant the best chance to win? Explain

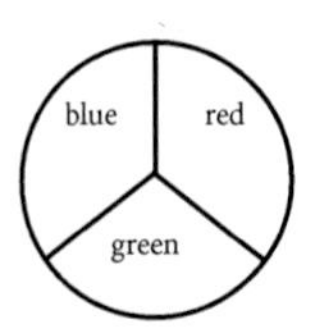

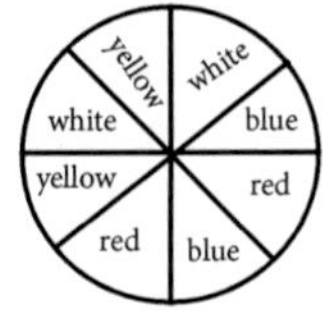

Spinner A Spinner B

2) You are going to do two separate events. First, you are going to roll a number cube with sides numbered 1 to 6. Second, you are going to spin a spinner with 4 equal parts (red, blue, green, & yellow). Put these four occurrences in order from least likely to most likely to happen on the number line. (use the letters A - D)

A) Rolling an even number B) Spinning a yellow
C) Spinning a color Not blue D) Rolling a six

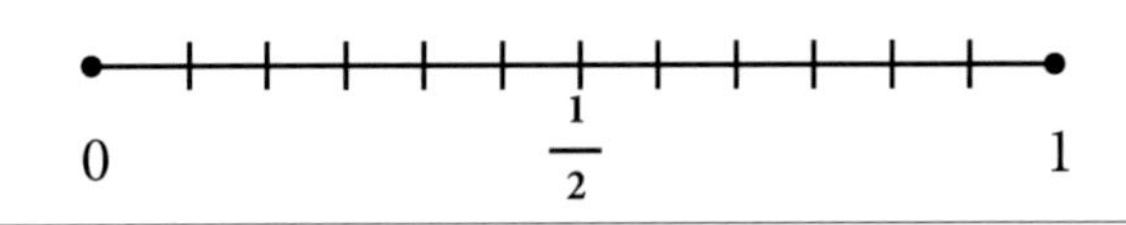

3) Below are the results of rolling a number cube (1-6) thirty times.

1111111 22222 33333333 44 55555 666

a) What is the experimental probability of getting a "6"? b) How is it different than the theoretical probability?

4) Below are the results of flipping a coin 16 times:

H H T H H T H H T H T H H H T T

How are the Experimental and Theoretical probabilities different for getting heads?

5) There are 25 m & m's in a bag. 15 are green. What is the probability that the first two eaten will be green?

6) There are 5 boys and 5 girls in Mr. Boland's homeroom. Two of the students are going to be randomly selected from the class. What is the probability that the two selected will both be girls?

7) You are one of 100,000 fans at a University of Michigan football game. The university is giving away a free car to one fan. The probability that you will be that fan would be located where on a number line?

a) Near zero

b) Near 1/4

c) Near 1/2

d) Near 3/4

e) Near 1

8) Mr. Anderson has 10 students in his homeroom (7 girls & 3 boys). Mrs. Brown has 10 students in her homeroom (4 girls & 6 boys). Each teacher randomly selects one student to pick up papers from the office. What is the probability that both teachers will send girls?

9) You are rolling a 12 sided die (numbers 1 to 12) 180 times. How many times would you expect to roll a "10"?

10) At a local coffee shop they offer 3 sizes (small, medium, & large) and 2 types of drinks (coffee & tea). Make a tree diagram showing all the possible outcomes.

11) You have a spinner with 5 equal sections (numbered 1 to 5). Represent your probability of rolling a "3" two times in a row as a: decimal, fraction & percent.

12) The numbers 1 to 10 are placed into a cup. What is the probability that 2 straight odd numbers will be selected. The first number is NOT replaced.

Name: ______________________________ Period: ____ ***Worksheet:***

1) What is the probability of spinning a number less than four on the spinner below?

2	4
6	8

2) Create a Frequency Table showing all the possibilities of flipping a coin (heads or tails) and rolling a number cube (1 to 6).

3) Roger rolled a six sided die (1-6) 90 times. Below are the results:

#1 = 20 #4 = 10
#2 = 15 #5 = 12
#3 = 18 #6 = 15

How are the experimental probabilities different than the theoretical probabilities?

4) A baseball team has 3 outfielders on a team of 9 players. If a player strikes out, what is the probability that the player is an outfielder?

5) You have a six sided number cube with numbers 1 to 6. What is the probability of rolling a number less than 6?

6) You are going to flip a coin 128 times. Based on the theoretical probability, how many times would you expect to flip a tail?

7) You are playing a game with an 8 sided die (numbers 1 to 8). What is the probability you get a number less than 4?

8) Steve is preparing to select a spot on the board below. What is the probability he will select a "B"?

A	B	A	C
D	A	B	C
A	C	E	B
A	C	A	D

9) Zelda has 2 types of shoes (sandals & tennis shoes), 2 types of shorts (nylon & cargo) and 3 types of shirts (rugby, polo, & fleece). Make an organized list of all the combinations.

10) Create a Frequency Table showing all the possible outcomes of selecting a beverage (pop, coffee, or tea) and a snack (grapes, apple, or banana).

11)What is the probability you could correctly guess the first 5 questions of a True and False quiz?

12) Is this an example of uniformed probability model? Explain.

A	B	C	D

ANSWER KEY *WORKSHEET #1*

1)

1	3	2
2	1	3
2	3	1

Archie can select any spot on the board. What is the probability he will select a 3? Is this an example of uniform probability?

Yes, all sections the same size

$3/9 = 1/3$

2) Ms. Fogler surveyed 30 7th grade students about their favorite colors. The following are the results: Blue (8), Red (7), Purple (10), and Yellow (5). If there are 180 students in the 7th grade, make a prediction (based on survey) how many would select each color.

Blue: 48

Red: 42

Purple: 60

Yellow: 30

3) A spinner is divided into 5 equal sections. Two sections are blue, one green, one yellow and one red.

What is the probability of ***NOT*** landing on red when spinning?

$4/5$

4) What is the probability that you could correctly guess the first four questions correct of a true & false quiz?

$1/16$

5) There are 30 students in your 5th period science class. Sixteen are girls. During the class, the office called for a student. The probability that the student was a girl would fall where on a number line?

a) zero

b) near zero

c) near middle

d) near one

e) one

6) A box contains 4 jelly filled and 6 cake doughnuts. What is the probability that the first 3 doughnuts eaten will be jelly filled?

$24/720 = 2/60 = 1/30$

7)

A	B
D	C

Tim spins the spinner above 20 times. These are his results:

A = 7	B = 5
C = 4	D = 4

How are these results different than the theoretical probability?

A = Occurred 2 more
C & D = Occurred 1 less
B = What was expected

8) What is the probability of rolling a "1" three straight times with a six sided number cube?

$1/216$

9) You are playing a game with the two spinners below. To win you must correctly select the letter and number.

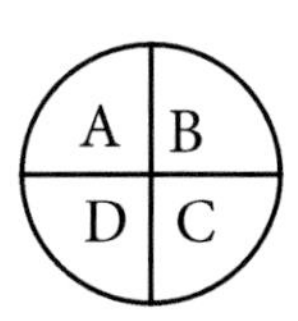

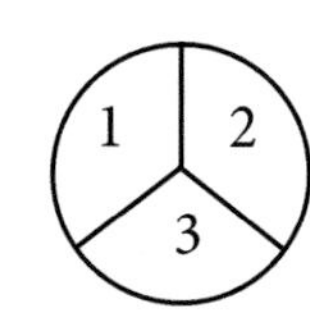

What is the probability of spinning the combination of "A" and "2"?

$1/12$

10) A restaurant offers 4 types of sandwiches (fish, hot dog, hamburger, grilled cheese) and three side dishes (fries, cole slaw, and onion rings). Create an organized list of all the possible outcomes if you order one sandwich and one side dish.

Fish, Fries
Fish, Cole Slaw
Fish, Onion Rings

Hamburger, Fries
Hamburger, Cole Slaw
Hamburger, Onion Rings

Hot Dog, Fries
Hot Dog, Cole Slaw
Hot Dog, Onion Rings

Grilled Cheese, Fries
Grilled Cheese, Cole Slaw
Grilled Cheese, Onion Rings

11) Is this an example of a uniformed probability model? Explain.

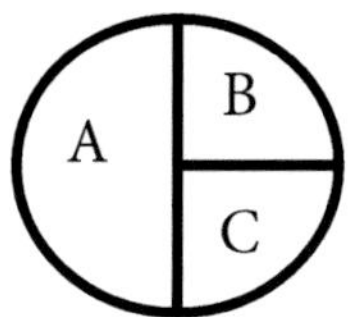

No.

A is larger than sections B & C

12) You flip a coin and roll a number cube (1 to 6). What is the probability that you will get the combination of tails and an even number?

$1/4$

ANSWER KEY *WORKSHEET #2* ***Worksheet:***

1) Mrs. Johnson's candy jar has 15 pieces of candy (7 orange, 5 grape, and 3 strawberry). Mrs. Johnson is letting four students select a piece of candy which they get to keep. What is the probability that each of the four pieces of candy will be strawberry?

$\frac{3}{15} \times \frac{2}{14} \times \frac{1}{13} \times \frac{0}{12} =$

$\frac{0}{32760} = 0$

2) Below is a different type of dart board for a new game. Is this an example of uniform probability? Explain.

B / C	A
A	D

No.
Section A = ½, D = ¼, C = ⅙, and B = 1/12
All sections should be same size

3) Tim is going to put the following 6 marbles into a ja
(a) What is the probability of drawing a white marble?
(b) Is this an example of uniformed probability?

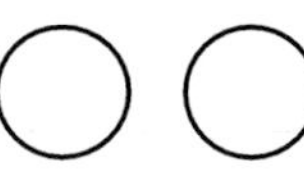
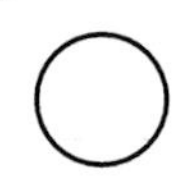

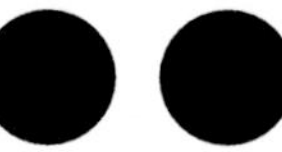

a) 3/6 = ½
b) No. Amounts different

4) A jar contains 7 blue, 5 yellow, 7 white and 15 red marbles. Are you more likely to pull out a blue or a yellow marble? Explain

Blue more likely
7/34 (Blue) vs. 5/34 (Yellow)

5) There are 20 jelly beans in a bag (5 yellow, 4 brown, 3 orange & 8 blue). What is the probability that the first three jelly beans you randomly eat will all be orange?

$\frac{3}{20} \times \frac{2}{19} \times \frac{1}{18} =$

$= \frac{6}{6840}$

$= \frac{3}{3420}$

$= \frac{1}{1140}$

6) Edward has two hats (visor & baseball) and three t-shirts (green, red, & yellow). Create a Frequency table showing all the possible combinations.

	G	R	Y
V	VG	VR	VY
B	BG	BR	BY

7) You are rolling an 8 sided die (numbers 1 to 8). What is the probability of rolling a number 7 or larger?

$\frac{2}{8} = \frac{1}{4}$

8) You are going to do two separate events. The first event is flipping a coin. The second event is spinning a spinner with three equal sections (Red, Blue, & Green). Put the probability of each event occurring in order from least likely to most likely. **a) Getting heads, b) Spinning a green, c) Spinning an orange, d) getting either heads or tails.**

1) Getting Orange (0/3) ***Least Likely***
2)Spinning Green (⅓)
3) Heads (½)
4) Getting either Heads or Tails (2/2 = 1) ***Most Likely***

9) There are 4 multiple choice questions (each with 4 choices - only one correct). You had to guess on all four of these questions. What is the probability that you were able to guess all four questions correctly?

$\frac{1}{4} \times \frac{1}{4} \times \frac{1}{4} \times \frac{1}{4}$

$= \frac{1}{256}$

10) A jar contains 7 blue, 5 purple, 3 white and 15 red marbles. What is the probably of selecting a blue marble?

$\frac{7}{30}$

11) You are going to do two separate events. First, you are going to roll a number cube with sides numbered 1 to 6. Second, you are going to flip a coin. Put these four occurrences in order from least likely to most likely to happen on the number line. (use the letters A - D)

A) Getting Heads B) Getting Heads or Tails
C) Getting a "1" D) Not Getting a "1"

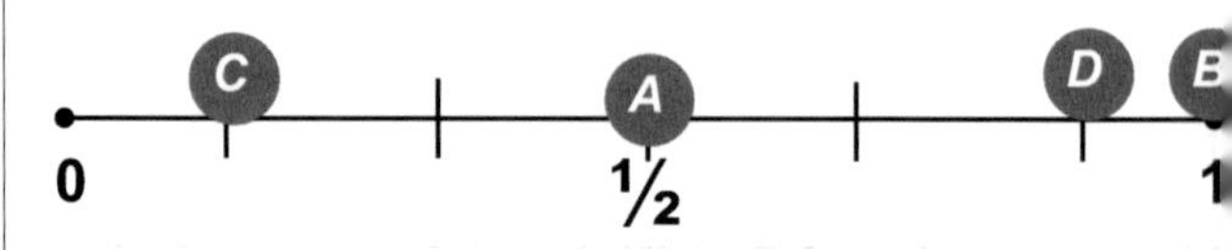

12) You have a six sided number cube (numbered 1 to 6). What is the probability of rolling an even number?

$\frac{1}{2}$

ANSWER KEY *WORKSHEET #3*

Worksheet: #3

1) You flip a coin and spin a spinner with four equal sections (sides A, B, C, D).

What is probability of flipping heads and spinning "B"?

You can represent the probability as either a fraction, decimal, or percent.

$\frac{1}{8}$, 0.125, 12.5%

2) Of the 20 students in Mr. Karp's science class, four are boys. If the office calls for a student, what are the chances it will be a girl?

A) Impossible

B) Unlikely

C) Likely

D) Certain

3) You have a spinner with three equal sections (A, B, C). Predict the number of times the spinner will land on "A" if you spin 100 times.

$33.\overline{33}$

33 or 34

4) There are 7 boys in a class of 15. What is the probability of the teacher selecting a girl to answer the first question?

$\frac{8}{15}$

5) You are rolling a number cube (numbers 1 to 6) fifty times. Predict how many times a number less than four will occur. Justify your answer.

$\frac{3}{6} = \frac{1}{2}$

25

6) You have a spinner with four equal sections numbered 1 to 4. What is the probability of spinning a number greater than 3?

$\frac{1}{4}$

7) Jenny has 3 shirts (green, purple, red) and 3 shorts (tan, black, khaki) to pick out an outfit for school.
Create a tree diagram showing all possible combinations.

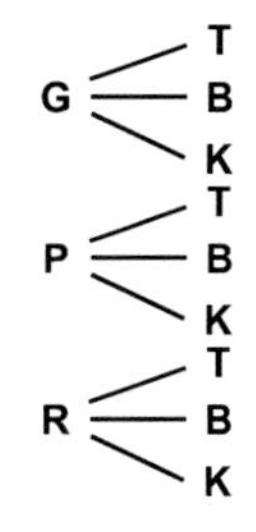

8) A teacher needs to select a boy & girl to go to the office. There are three boys (Tim, Sam, Jack) and two girls (Mary, Valerie). Create an Organized List of all the outcomes.

	T	S	J
V	VT	VS	VJ
M	MT	MS	MJ

9) Represent the probability of getting heads, when flipping a coin two times in a row as a decimal, fraction, and percent %.

$\frac{1}{4}$, 0.25, 25%

10) You are rolling a number cube (with the numbers: 1, 2, 3, 4, 5, 6) 120 times. Predict how many times the number 4 will occur. (theoretical probability).

20

11) You are flipping a coin 44 times. Based on the Theoretical Probability, predict the number of times tails will occur.

22

12) There are 13 red marbles, 5 green marbles and 2 blue marbles in a jar. Represent the probability of selecting a red marble as a fraction, decimal, and percent.

$\frac{13}{20}$, 0.65, 65%

ANSWER KEY *WORKSHEET #4* **Worksheet:**

1) If you flip a coin three times what is the probability of getting heads all three times? (write as a fraction)

$^1/_8$

2) You have a spinner with four equal sections (A, B, C, D). If you spin the spinner 80 times, how many times would you expect the "B" to occur?

20

3) There are 30 major league baseball teams. One is from Canada (Toronto Blue Jays). If each team had equal talent, how would you describe Toronto's chance of winning the World Series next year? (circle)

Impossible

Unlikely

Likely

Certain

4) If you flip a coin and roll a number cube (1 to 6) what is the probability you will get the combination of "Heads" and the number "5"?

$^1/_{12}$

5) Is this an example of uniform probability model?

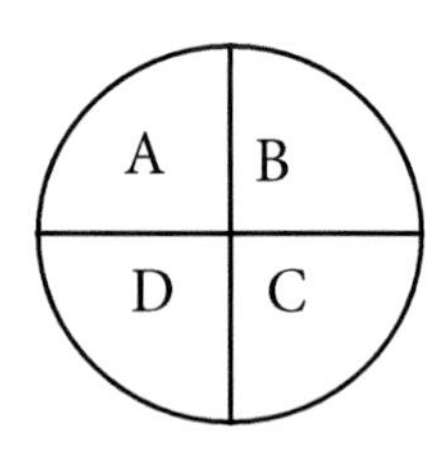

Yes, all four sections are equal.

6) There are 15 blue cars and 5 red cars in the parking lot. Based on this sample, predict how many of the next 8 cars will be blue and red.

$^{15}/_{20}$ = ¾ Blue
= ¼ Red

6 Blue
2 Red

7) At the local coffee shop there are 3 sizes (S, M, L) and 4 drinks (Espresso, Frappuccino, Coffee, & Tea). Make a tree diagram of the different possible outcomes.

8) The following are the results of flipping a coin 20 times.

T H H H H T H H T T
T H H H T H H H T H

What is the experimental probability of tails?

$^7/_{20}$

9) You are one of the 200,000,000 people playing the lottery Friday. Your probability of winning would be located where on a number line?

a) Near zero

b) Near the middle

c) Near 1

10) There are 3 boys & 2 girls in Mr. Smith's class. There is 1 boy & 4 girls in Mrs. Greene's class. What is the probability a boy from Mr. Smith's class and Mrs. Greene's class will be selected as class representatives? Write as a fraction or percent.

$^3/_{25}$ = 12%

11) There are 11 players on a football team. Five of the players are offensive linemen. If there is a penalty on the offense what is the probability it is on one of the offensive linemen?

$^5/_{11}$

12) You are playing cards with a standard deck of cards (52). In the first 40 games, how many times would you predict a face card (Jack, Queen, or King) would be the first card delt?

A face card would occur first 9.23 times out of 40. So...

9 or 10

ANSWER KEY *WORKSHEET #5*

1) A local radio station is having a contest to win a car. To win, a person must spin a spinner and land on a ***BLUE*** space. Each contestant has the choice of two spinners (below)to select from. Which spinner gives the contestant the best chance to win? Explain

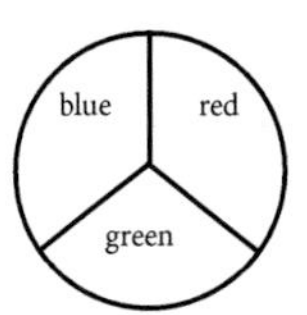

Spinner A

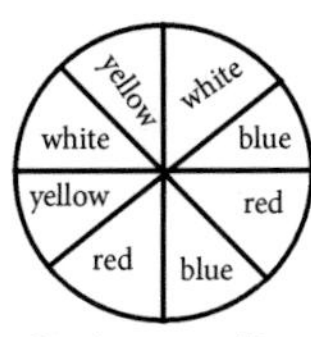

Spinner B

⅓ (33%) Spinner A

¼ (25%) Spinner B

Spinner A is better

2) You are going to do two separate events. First, you are going to roll a number cube with sides numbered 1 to 6. Second, you are going to spin a spinner with 4 equal parts (red, blue, green, & yellow). Put these four occurrences in order from least likely to most likely to happen on the number line. (use the letters A - D)

A) Rolling an even number ½ B) Spinning a yellow ¼

C) Spinning a color Not blue ¾ D) Rolling a six ⅙

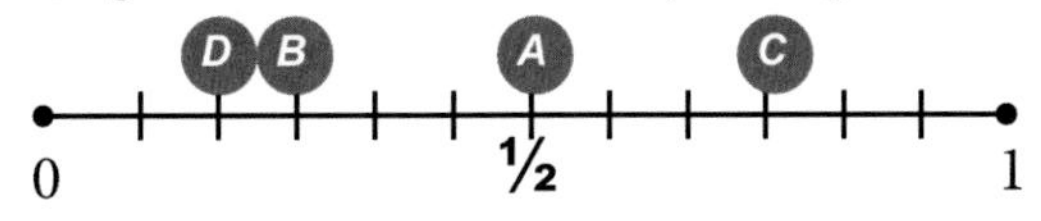

3) Below are the results of rolling a number cube (1-6) thirty times.

1111111 22222 33333333 44 55555 666

a) What is the experimental probability of getting a "6"? b) How is it different than the theoretical probability?

$3/30 = 1/10$ (10%) Experimental Probability

$5/30 = 1/6$ (16.7%) Theoretical Probability

The "6" happened 2 less than expected

4) Below are the results of flipping a coin 16 times:

H H T H H T H H T H T H H H T T

How are the Experimental and Theoretical probabilities different for getting heads?

$10/16 = 5/8$ (62.5%) Experimental Probability

$8/16 = 1/2$ (50%) Theoretical Probability

"Heads" happened 2 more than expected

5) There are 25 m & m's in a bag. 15 are green. What is the probability that the first two eaten will be green?

$15/25 \times 14/24 = 210/600 = 7/20$

6) There are 5 boys and 5 girls in Mr. Boland's homeroom. Two of the students are going to be randomly selected from the class. What is the probability that the two selected will both be girls?

$5/10 \times 4/9 = 20/90 = 2/9$

7) You are one of 100,000 fans at a University of Michigan football game. The university is giving away a free car to one fan. The probability that you will be that fan would be located where on a number line?

a) Near zero

b) Near 1/4

c) Near 1/2

d) Near 3/4

e) Near 1

8) Mr. Anderson has 10 students in his homeroom (7 girls & 3 boys). Mrs. Brown has 10 students in her homeroom (4 girls & 6 boys). Each teacher randomly selects one student to pick up papers from the office. What is the probability that both teachers will send girls?

$28/100 = 7/25$

9) You are rolling a 12 sided die (numbers 1 to 12) 180 times. How many times would you expect to roll a "10"?

15

10) At a local coffee shop they offer 3 sizes (small, medium, & large) and 2 types of drinks (coffee & tea). Make a tree diagram showing all the possible outcomes.

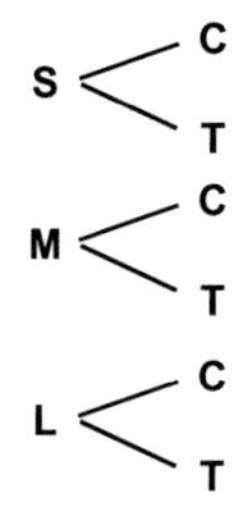

11) You have a spinner with 5 equal sections (numbered 1 to 5). Represent your probability of rolling a "3" two times in a row as a: decimal, fraction & percent.

$1/25$

.04

4%

12) The numbers 1 to 10 are placed into a cup. What is the probability that 2 straight odd numbers will be selected. The first number is NOT replaced.

$20/90 = 2/9$

ANSWER KEY *WORKSHEET #6*

Worksheet:

1) What is the probability of spinning a number less than four on the spinner below?

2	4
6	8

$\frac{1}{4}$

2) Create a Frequency Table showing all the possibilities of flipping a coin (heads or tails) and rolling a number cube (1 to 6).

	1	2	3	4	5	6
H	H1	H2	H3	H4	H5	H6
T	T1	T2	T3	T4	T5	T6

3) Roger rolled a six sided die (1-6) 90 times. Below are the results:

#1 = 20 #4 = 10
#2 = 15 #5 = 12
#3 = 18 #6 = 15

How are the experimental probabilities different than the theoretical probabilities?

The Theoretical Probability is15 times for each number
#1 - 5 More
#3 - 3 More
#4 - 5 Less
#5 - 3 Less
#2 & #6 - Same

4) A baseball team has 3 outfielders on a team of 9 players. If a player strikes out, what is the probability that the player is an outfielder?

$\frac{1}{3}$

5) You have a six sided number cube with numbers 1 to 6. What is the probability of rolling a number less than 6?

$\frac{5}{6}$

6) You are going to flip a coin 128 times. Based on the theoretical probability, how many times would you expect to flip a tail?

64

7) You are playing a game with an 8 sided die (numbers 1 to 8). What is the probability you get a number less than 4?

$\frac{3}{8}$

8) Steve is preparing to select a spot on the board below. What is the probability he will select a "B"?

A	B	A	C
D	A	B	C
A	C	E	B
A	C	A	D

$\frac{3}{16}$

9) Zelda has 2 types of shoes (sandals & tennis shoes), 2 types of shorts (nylon & cargo) and 3 types of shirts (rugby, polo, & fleece). Make an organized list of all the combinations.

Sandals, Nylon, Rugby
Sandals, Nylon, Polo
Sandals, Nylon, Fleece

Sandals, Cargo, Rugby
Sandals, Cargo, Polo
Sandals, Cargo, Fleece

Tennis shoe, Nylon, Rugby
Tennis shoe, Nylon, Polo
Tennis shoe, Nylon, Fleece

Tennis shoe, Cargo, Rugby
Tennis shoe, Cargo, Polo
Tennis shoe, Cargo, Fleece

10) Create a Frequency Table showing all the possible outcomes of selecting a beverage (pop, coffee, or tea) and a snack (grapes, apple, or banana).

	P	C	T
G	GP	GC	GT
A	AP	AC	AT
B	BP	BC	BT

11)What is the probability you could correctly guess the first 5 questions of a True and False quiz?

$\frac{1}{32}$

12) Is this an example of uniformed probability model? Explain.

A	B	C	D

Yes, all four sections are equal.

The following are six problems can be used as exit tickets or just as practice to show work

60 students at Anderson Middle School each flipped a coin. 40 of the students coin flips resulted in "Heads." Compare what you thought should happen (Theoretical Probability) with the actual results (Experimental Probability).

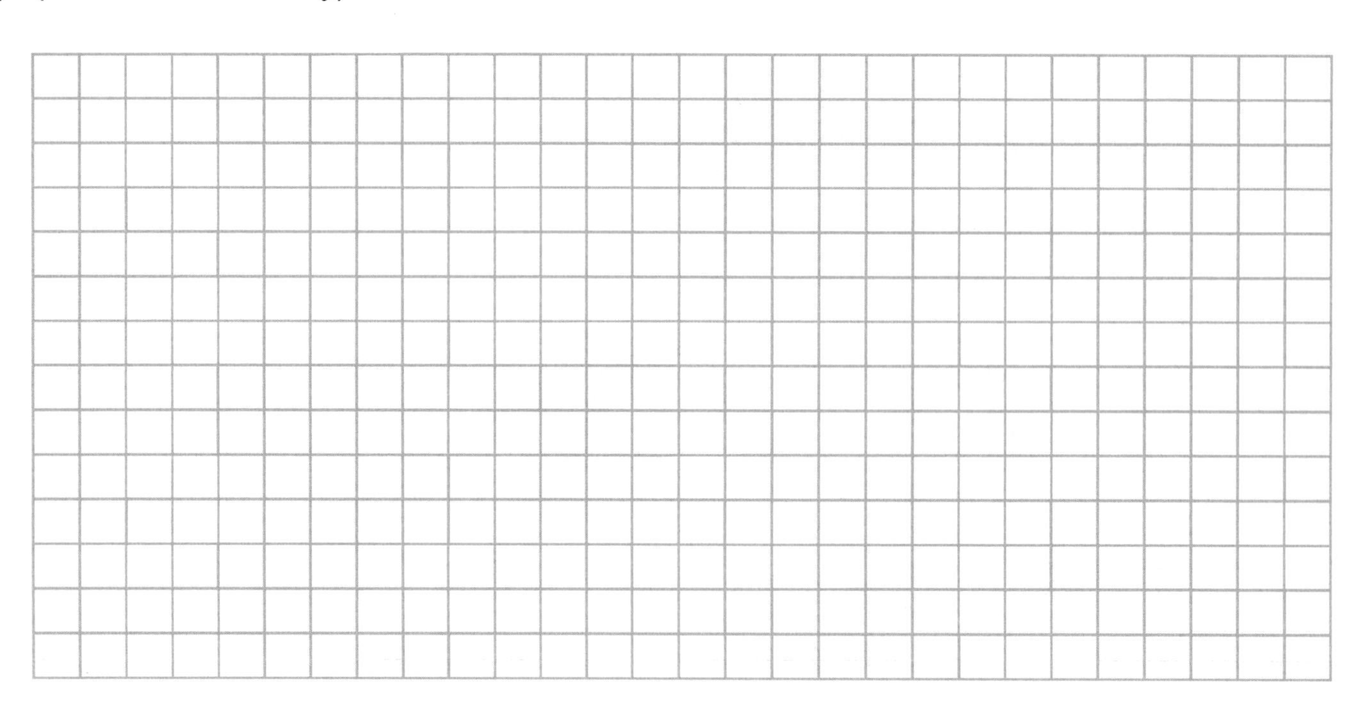

In Mr. Jenkins math classes, 48 students rolled a number cube (number from 1 to 6). How many students would you predict rolled a "2"? Show how you calculated your answer.

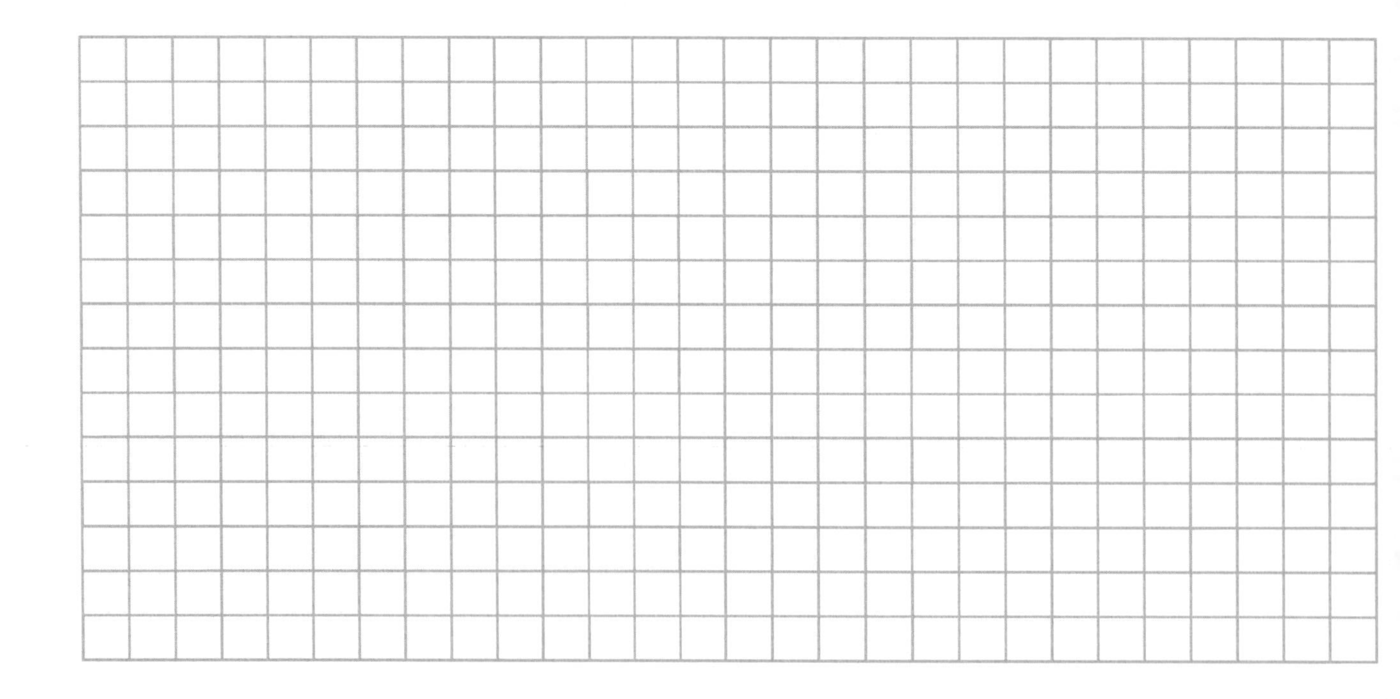

#3

arcy is playing a game in which she rolls two number cubes (each numbered 1-6). On her last roll he needs to roll a sum of greater than "9" to win. What is her probability of winning? Show work.

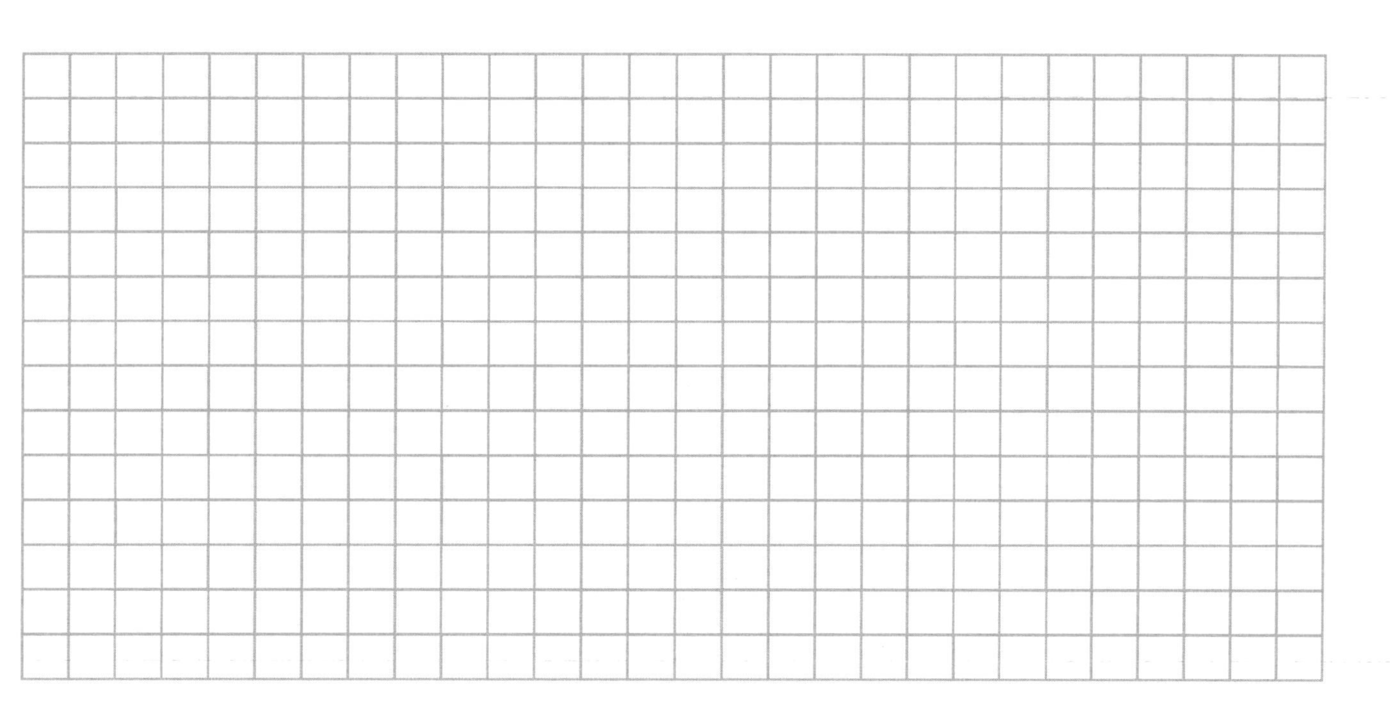

#4

here are 3 green and 2 blue marbles in a cup. Scott and Kevin are debating the odds of randomly electing 3 straight green marbles. Scott thinks the best strategy is to select a marble and return it to e cup before making the next selection. Kevin thinks the best strategy is to keep the marble after ach selection. Which strategy gives you the best probability of picking 3 straight green marbles?

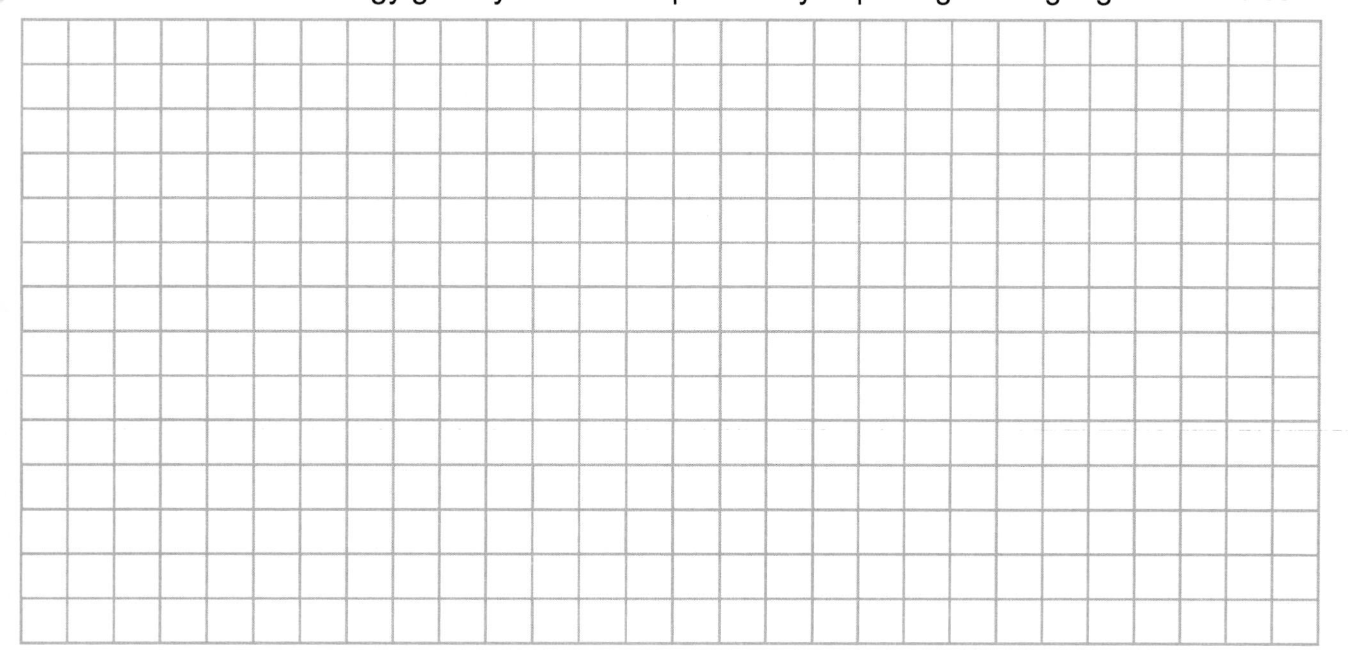

B | A
C |
A | D

A gym class at Plainfield Elementary School uses the board to the left when playing a game with bean bags. 800 students randomly tossed bean bags at the board. Based on the layout of the board predict how many of the 800 bean bags landed in section "C" of the board. Show work.

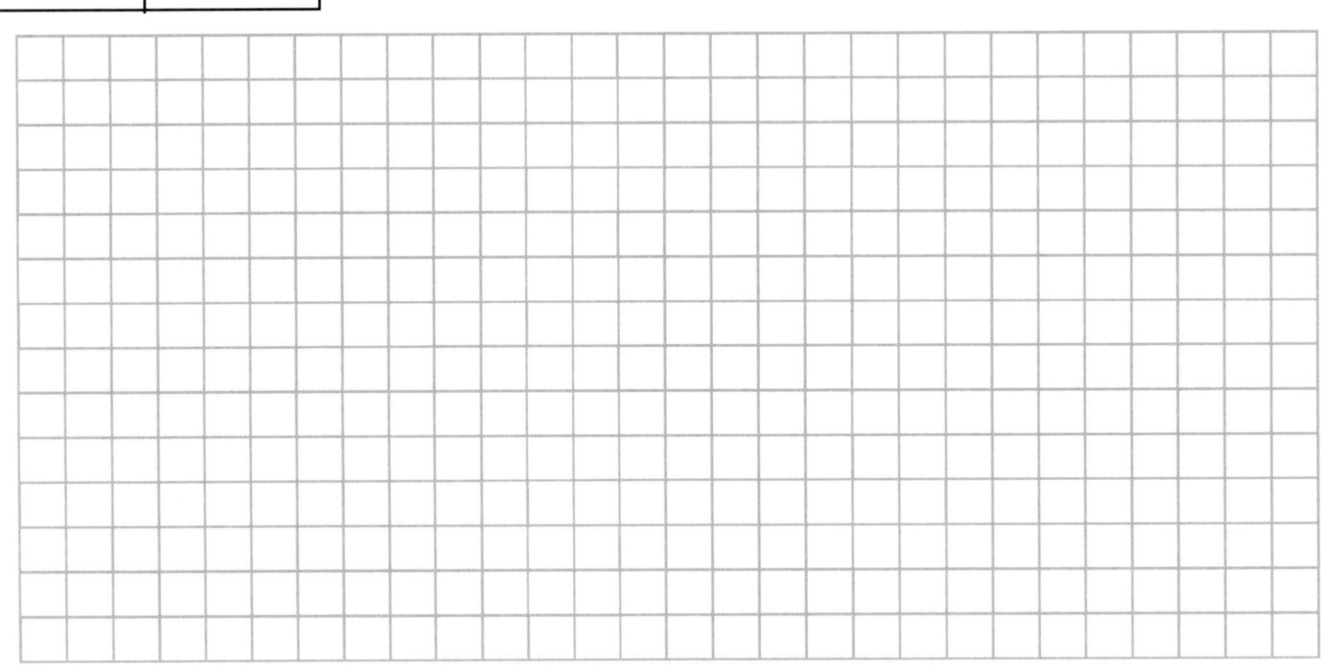

#6

The Paradise Valley baseball team has three of the leagues best home run hitters. Derek hits a hom run 5% of the time, Tino hits a home run 10% of the time and Reggie hits a home run 20% of the time. These three players are the first three to bat next inning. What is the probability (as a fraction c decimal) that the three players will hit 3 consecutive home runs?

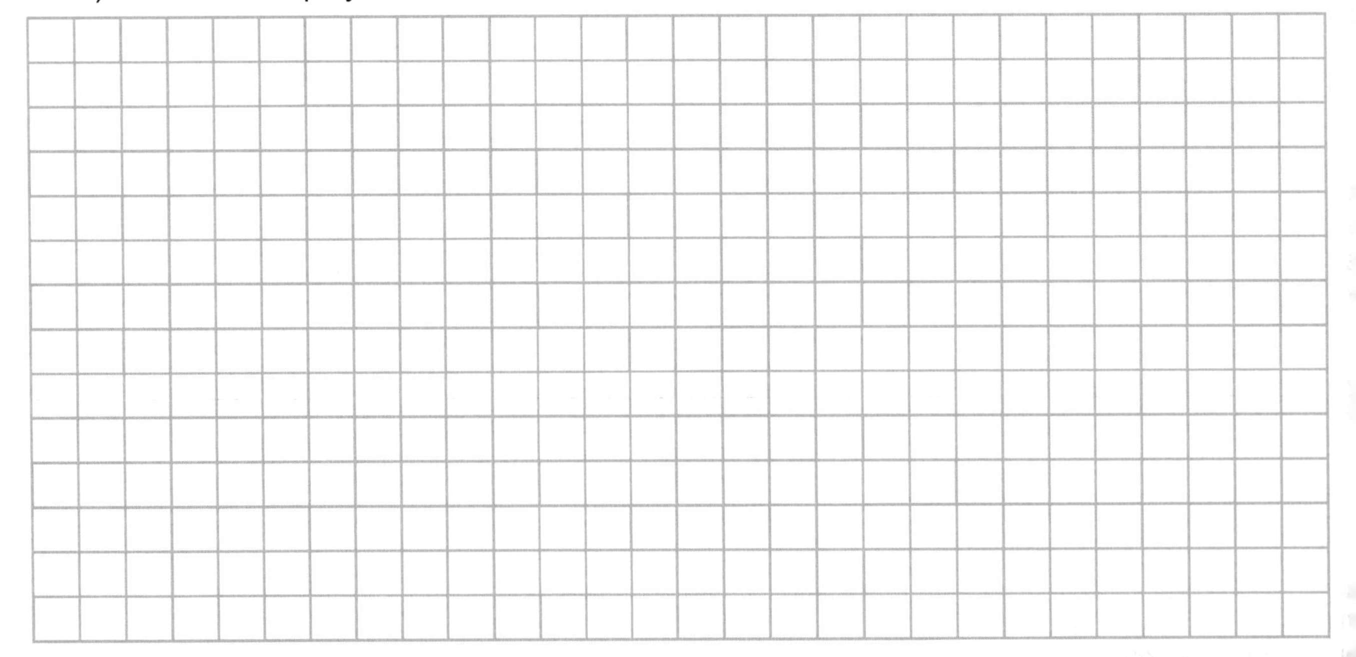

Key for Short Answers

1. The Probability of getting Heads is 1/2 or 50%. Based on this, you would expect (or predict) that half of the students will get heads when they flip their coin. Half (50%) of 60 is 30 so we would expect 30 students to get Heads. When the students actually flipped the coins Heads occurred 40 times. This is 10 more than expected.

2. The Probability of getting a "2" when rolling a number cube is 1 out of 6 ($\frac{1}{6}$). Each of the numbers (1,2,3,4,5, & 6) should occur the same amount. To find $\frac{1}{6}$ of 48 when can take 48 and divide by six which equals 8 ($48 \div 6 = 8$). We could also take 48 and multiply it by the probability ($\frac{1}{6}$) and get 8 ($48 \times 1/6 = 8$). So there should be 8 students that get the number "2."

3. There are 36 different combinations ($6 \times 6 = 36$) or write all out as an organized list or Frequency table. Of the 36 different combinations, 6 of them have sums greater than 9. So there is a 6 out of 36 ($\frac{6}{36}$)chance of getting a sum greater than 9. $\frac{6}{36}$ will simplify to $\frac{1}{6}$. So she has a $\frac{1}{6}$ chance of winning the game.

1 + 1 = 2	2 + 1 = 3	3 + 1 = 4	4 + 1 = 5	5 + 1 = 6	6 + 1 = 7
1 + 2 = 3	2 + 2 = 4	3 + 2 = 5	4 + 2 = 6	5 + 2 = 7	6 + 2 = 8
1 + 3 = 4	2 + 3 = 5	3 + 3 = 6	4 + 3 = 7	5 + 3 = 8	6 + 3 = 9
1 + 4 = 5	2 + 4 = 6	3 + 4 = 7	4 + 4 = 8	5 + 4 = 9	*6 + 4 = 10*
1 + 5 = 6	2 + 5 = 7	3 + 5 = 8	4 + 5 = 9	*5 + 5 = 10*	*6 + 5 = 11*
1 + 6 = 7	2 + 6 = 8	3 + 6 = 9	*4 + 6 = 10*	*5 + 6 = 11*	*6 + 6 = 12*

Getting 10
6 + 4 = 10
4 + 6 = 10
5 + 5 = 10

Getting 11
6 + 5 = 10
5 + 6 = 10

Getting 12
6 + 6 = 12

	1	2	3	4	5	6
1	2	3	4	5	6	7
2	3	4	5	6	7	8
3	4	5	6	7	8	9
4	5	6	7	8	9	*10*
5	6	7	8	9	*10*	*11*
6	7	8	9	*10*	*11*	*12*

4. ***Scott*** would have a $\frac{3}{5}$ chance (60% chance) with each selection. So, $\frac{3}{5} \times \frac{3}{5} \times \frac{3}{5} = \frac{27}{125}$ which equals 21.6%. Or you could multiply percents as a decimal $0.6 \times 0.6 \times 0.6 = .216 = 21.6\%$

Kevin chance would differ each time as marbles are selected. So, $\frac{3}{5} \times \frac{2}{4} \times \frac{1}{3} = \frac{6}{60} = \frac{1}{10}$ which equals 10%. Or you could multiply percents as a decimal $0.6 \times 0.5 \times 0.33$ (rounded) $= .099 = 10\%$

Scott has more than double the chance of winning compared to Kevin.

5. One way is to visualize this board as 4 equal sections (A, A, D & a combo of B/C). Take the total of 800 and divide by 4 which equals 200 ($800 \div 4 = 200$). So each section equals 200 tosses. To find the amount for "C" divide 200 by 2 which equals 100 ($200 \div 2 = 100$). So the Theoretical Probability of "C" occurring is 100 times out of 800.

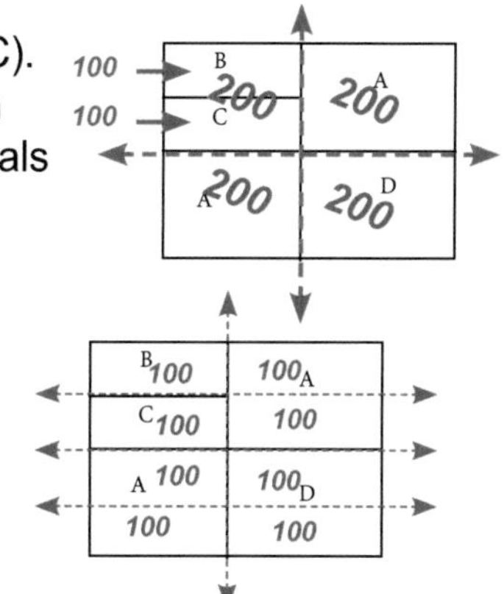

nother way to visualize this board is as 8 equal sections (Split A, A, & D in half). ake the total of 800 and divide by 8 which equals 100 ($800 \div 8 = 100$). So each ection equals 100 tosses. "C" equals one of these sections so the Theoretical robability of "C" occurring is 100 times out of 800.

6. There are three events (the three players batting). You can solve using fractions and/or decimals. **Using Fractions**: Derek hits a home run 5% of the time which is $\frac{1}{20}$ ($\frac{5}{100} = \frac{1}{20}$), Tino hits a home run 10% of the time which is $\frac{1}{10}$ ($\frac{10}{100} = \frac{1}{10}$) and Reggie hits a home run 20% of the time which is $\frac{1}{5}$ ($\frac{20}{100} = \frac{1}{5}$). If you multiply their chances of occurring you find that the probability of hitting three consecutive home runs is $\frac{1}{1000}$. $\frac{1}{20} \times \frac{1}{10} \times \frac{1}{5} = \frac{1}{1000}$

sing Decimals: Derek hits a home run 5% of the time which is .05, Tino hits a home run 10% of the time hich is .10 & Reggie hits a home run 20% of the time which is .20. If you multiply $(.05) \times (.10) \times (.20) =$ **.001** u find the solution which is the same as $\frac{1}{1000}$.

Made in the USA
Coppell, TX
30 May 2022